Please return/renew this item
by the last date shown.
Books may also be renewed by
phone and Internet.

contents

meet the author

Welcome to *Plumbing Made Easy*!

Strange as it may sound, I was in the last generation of children to leave school at the age of 14. This makes me sound old, but no, I'm only in my early 50s! I first started teaching full time in 1985 and in that time it seems that the world of plumbing has gone through a 'hyper leap'. For many years things just seemed to stay the same and any changes made seemed to be a slow, steady progression. Joints to lead and cast-iron pipes were things that had to be learned, and they were skills that made you feel a real specialist in your craft.

Today these skills are no longer required and things are put together with such speed – plastic pipes and their push-fit joints are used for all kinds of installations. Legislation today, however, takes a much more prominent position. When I first started trading as a plumber, you could, within reason, do as you pleased to make a system work: gas work, boiler installation, drainage alterations, etc. Today, this is not the case: gas engineers must be registered and can only undertake work in the areas in which they have been assessed. New boilers must meet minimum standards of energy-efficiency, and installation must be registered with the local authority. If you alter any plumbing or electrical installations, approval will be required in order to comply with the law.

So, while in one way plumbing has become easier, in another it has become very much harder if you want to trade. Undertaking plumbing yourself is something that many can do, but remember that there is legislation that must be abided by. Moreover, if you install a system incorrectly, it may only work for a limited lifespan. Take your time to fully appreciate how the system works and, if in doubt, seek expert advice.

Good luck!

Roy Treloar

introduction

This book has been written with the domestic homeowner in mind. It identifies the plumbing systems in your home. You might be considering the replacement of extensive pipework, or perhaps you'd just like to know what to do in an emergency. In any case, this book will give you insight into the activities that a plumber might undertake should they be called upon and it provides you with clues about what they might need to do and why. It will also provide you with key questions to ask when seeking the services of a plumber.

This book gives you information about the water supply pipe: how it enters the building from the street outside and travels through your house. Dealing first with the cold water supply, then the hot, it will show you how the heating works and also teaches you about the drainage of water from your property. As the book takes you around your home it identifies the main variations of plumbing systems found, and shows you specific things to look out for in the design, thus hopefully ensuring a trouble-free existence.

Hopefully you will be able to find the courage to try some of the smaller jobs yourself, and you might surprise yourself and gain enough confidence to tackle much bigger tasks in the fullness of time. With the escalating cost of calling in a plumber these days, you should get your money back on the first successfully completed activity. I hope that this book brings you some happy plumbing results!

1

the plumbing in your home

This first chapter looks at the plumbing system within your home. It takes you on the journey of the cold inlet supply pipe from the point where the water is fed to the property and building and passes through the pipework, to the point where it supplies its point of use, which may be directly to an appliance, such as a sink outlet, toilet or storage cistern that feeds a hot water or central heating system. The chapter also gives an overview of the different types of water, i.e. 'hard' or 'soft', and identifies the possible problems these water types may have on a system, as well as discussing ways in which to overcome such problems. Finally the chapter looks at the drainage of water from the building via the drainage pipes and the methods employed to ensure smells do not enter the building from the house drains and sewer.

Incoming cold water supply

The water pipe feeding into your home comes from a supply pipe in the road, at a point just outside your property. There is usually a water authority valve at this point, and it is here that your responsibility for the water and pipework begins. The pipe travels below ground at a minimum depth of 750 mm to ensure that it is protected from damage and that the water will not freeze if the temperature drops below freezing point (0°C). The pipe then passes into a pipe duct through the foundations and ground floor into your home, terminating with a stopcock (tap).

In newer buildings a water meter will be incorporated within this supply pipe. This may be contained within a chamber outside, keeping the meter below ground level, or within the building itself, thereby allowing easier access for reading and maintenance. There may also be a stopcock situated underground at the boundary to your property, in addition to the one inside.

The pipe in the road from which this drinking water supply is taken is usually referred to as the 'mains'.

The water supply pipe

For the past 30 years or so, plastic (polyethylene) has been used for the cold water pipe feeding your home. Today it is typically blue and the standard diameter is 25 mm (which is equivalent to a copper pipe of 22 mm diameter) and is adequate to supply several outlets at once. In the past, however, smaller-sized pipes were used, including:

* 20 mm plastic pipe – either black or blue (equivalent to 15 mm copper pipe size)
* 15 mm copper pipe
* ½" galvanized mild steel pipe
* ½" lead pipe.

These older pipes are regarded as too small in a modern house because of the extra appliances used (washing machine, showers, etc.) and extra toilets. The size can restrict the flow of

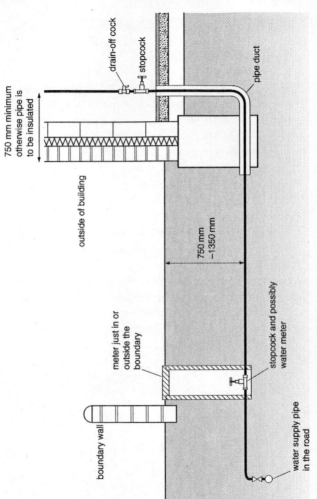

drain-off cock

stopcock

pipe duct

750 mm minimum otherwise pipe is to be insulated

outside of building

750 mm –1350 mm

meter just in or outside the boundary

stopcock and possibly water meter

boundary wall

water supply pipe in the road

Figure 1.1 *Cold water supply into a home.*

water and cause a loss of water flow at some outlets if several appliances are opened at the same time. Unfortunately there is not a lot you can do with your existing supply pipe if it's too small, other than replacing it with a new pipe.

Supply stopcock (stoptap)

It is very important that you know the location of this valve; after all, it supplies the water to the building, and turning it off will stop the flow of water. This is essential in a situation where water is leaking from pipework. Typical locations for the stopcock inside the building are:

* under the kitchen sink
* in a downstairs toilet
* under the stairs, in a cupboard
* in the garage
* in the basement
* under a wooden floorboard, just inside the front door.

There may be an additional stopcock outside the building. Don't turn off this valve until you fully understand the consequences of doing so; this will be discussed in Chapter 2.

Ideally, once the internal stopcock has been found, a label should be tied to the operating handle, so that anyone needing to find it in the future will know that this is the main water inlet to the building.

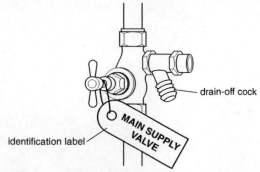

drain-off cock

identification label

MAIN SUPPLY VALVE

Figure 1.2 *Supply stopcock with drain-off valve.*

Cold supply inside the dwelling

Once you have identified the incoming supply, look for a small outlet valve, known as a drain-off cock, just after the stopcock or incorporated within its design. This may be missing in older buildings or in poorly installed systems. The drain-off cock allows the cold water supply mains pipework to be drained, for example for maintenance work or if you're going away for a long period of time in winter. There is the provision for a hose connection, but generally when the supply has been shut off, much of the water can be drained out via the kitchen sink, so that only that remaining in the pipe needs to be drained.

From the stopcock the pipe will run to the kitchen sink and other outlets. The route will depend on the system design, which will be one of the following:

* direct cold water supply
* indirect cold water supply
* modified cold water supply.

The pipework usually runs beneath floors or through pipe ducts, for example alongside the vertical soil or drainage stack (the drainpipe taking waste water from the building) as it passes up through the building. It may also be encased within the plaster wall. In all cases the actual pipe route is not a major concern provided that it is protected from unforeseen damage and frost.

Direct cold water supply

If you have this system, all of your cold water outlet points are fed directly from the mains supply. These include all appliances such as the sink, bath, basin and WC, plus any other outlets to washing machines, dishwashers or outside taps used for watering the garden. The cold supply may also feed a hot water system such as an unvented domestic hot water supply or combination boiler, discussed in Chapter 2.

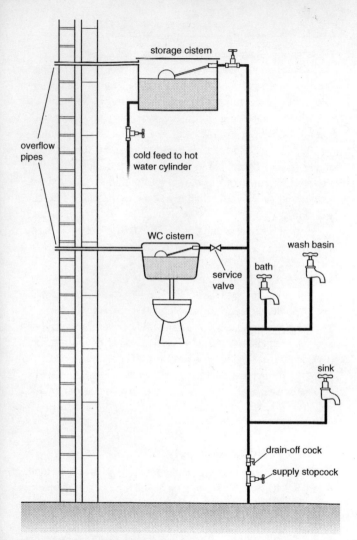

Figure 1.3 *Direct cold water supply system.*

Indirect cold water supply

In this system the only appliance fed directly from the mains supply is the kitchen sink, plus a water softener if one is incorporated within the property. Instead of feeding directly to the other appliances, the supply feeds a water storage cistern, usually found within the roof space (loft). All other outlet points in the building are then fed from this storage cistern (Figure 1.4).

Modified cold water supply

This type of system is a combination of both the direct and indirect supply systems. In other words, there may be several outlets from the mains supply and several fed via a storage cistern.

Prior to the 1980s most systems were of the indirect design. These were designed to maintain a flow of water under the worst possible conditions, for example when the supply was cut off for some reason, such as the water authority doing essential repairs, or in areas where there was an excessive drop in water pressure at peak times.

The local water authority may also have imposed a specific requirement that the supply had to be of an indirect design. However, today, due to higher pressures and consumer demand for combination boilers, unvented hot water supplies and guaranteed availability of drinking water, more and more systems rely on direct mains supply pipework. Also, with the direct system supplying both cold and hot water there is no need to have a cistern in the roof space or to extensively insulate the pipework and cistern from freezing up in the winter.

It is important to note that where all outlets are supplied via the mains supply, the supply pipe must be of a sufficient size (minimum 22 mm), otherwise, as mentioned earlier, some outlets will be starved of water when several outlets are open at the same time.

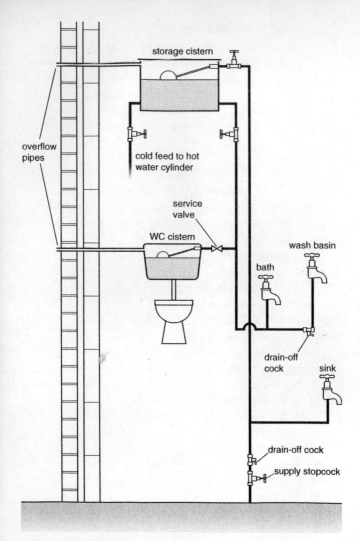

Figure 1.4 *Indirect cold water supply system.*

What outlets are fed directly from the supply main?

Finding out which outlets are fed directly from the cold mains supply pipe in your home is a simple process. First, turn off the incoming stopcock (as explained above) and then go around to all outlet points (taps) on the system to see which do not have any water flow available when the tap is turned on. Likewise, to find out if the toilet cistern is fed from the mains supply, flush the toilet to see if it refills.

Drinking water (potable water)

It may be a surprise to learn that if modern systems are designed and installed correctly, all outlet points, both hot and cold, should be supplied with water fit for human consumption, even where they are supplied via a cistern in the roof space. When we look at the installation of the pipework and appliances you will learn that the water must be protected from contamination at all costs. For example, in Figure 1.5 you will see that a filter has been incorporated within the overflow and that the cistern itself has a tight-fitting lid with all connections designed to prevent anything getting in and contaminating it, such as insects. So, water that has been stored in a cistern will also be regarded as safe to drink, and you must ensure, under all circumstances, that it remains this way.

The storage cistern

Figures 1.3 and Figure 1.4 show a cistern that contains a large volume of water for the purpose of supplying hot or cold water pipework that is not fed directly from the mains. Buildings in which everything is fed directly from the cold mains water supply do not have a storage cistern.

The water level inside the cistern is regulated by the use of a float-operated valve, designed to close off the water supply when the desired water level is reached. Should this valve fail to

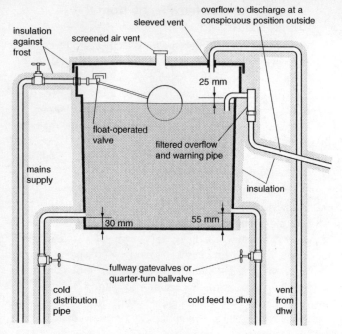

insulation against frost

screened air vent

sleeved vent

overflow to discharge at a conspicuous position outside

25 mm

float-operated valve

filtered overflow and warning pipe

mains supply

insulation

30 mm

55 mm

fullway gatevalves or quarter-turn ballvalve

cold distribution pipe

cold feed to dhw

vent from dhw

dhw = domestic hot water

Figure 1.5 *Cold water feed and storage cistern.*

operate, the water will continue to rise until the overflow pipe is reached, at which point it will overflow, warning the occupants of the building that something is wrong.

For the past 35 years or so, storage cisterns have been made of plastic materials, but some very old galvanized cisterns can still be found. Where this is the case it may be worth considering a replacement as it might have exceeded its expected lifespan.

All storage cisterns fitted since 1991 should be of a design that incorporates a tight-fitting lid and filtered overflow to ensure that even the smallest of insects cannot get in to contaminate the water supply. Even the vent pipe from the hot water supply (discussed later) passes through a rubber grommet in the lid.

Around all this is a snugly fitted insulation jacket, and all the pipework to and from the cistern should be similarly insulated. Older installations may not be protected to such a high standard, and if an inadequate system is encountered (for example, with a loose or flimsy lid) the water should be treated with caution where it is used at cold or hot water outlets. If there is no lid at all, this needs to be remedied immediately. Dead bats are commonly found floating and rotting in unprotected cisterns.

The condition of the storage cistern needs to be inspected occasionally to check that it is sound and protected. Ideally, once a year, remember to check:

* the filters found in the overflow and lid to ensure that they are not blocked, for example with flies
* the operation of the float-operated valve, to ensure that it is closing properly.

The float-operated valve (ballvalve)

The float-operated valve found within the storage cistern is generally of the same type as that found within a toilet cistern, although many of the newer toilet cistern float-operated valves are of a different design. The float-operated valve is often simply called a ballvalve, taking its name from the large ball, attached to the lever arm, which floats on the surface of the water. As the water inside the cistern rises and falls, so does the float.

These valves work on the principle of leverage, in that as the water rises the long arm lifts and forces a washer up against the water supply inlet. Generally only two designs of float-operated valve will be found, as shown in Figure 1.6. The older valve, known as the Portsmouth ballvalve, can no longer be installed as it contravenes current Water Supply Regulations. There are two main reasons for this:

1 Its inlet will be submerged at times when the valve is overflowing. If you look closely at the two valve designs, you will notice that the Portsmouth valve lets water into the cistern from below the valve body, whereas the diaphragm valve lets water into the cistern from above

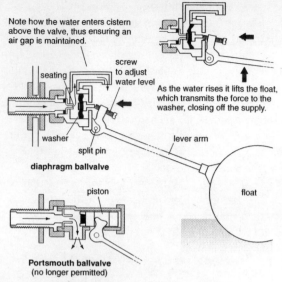

Note how the water enters cistern above the valve, thus ensuring an air gap is maintained.

seating

screw to adjust water level

As the water rises it lifts the float, which transmits the force to the washer, closing off the supply.

washer

split pin

diaphragm ballvalve

lever arm

piston

float

Portsmouth ballvalve
(no longer permitted)

Figure 1.6 *Float-operated valves (ballvalves).*

the valve body. The advantage of discharging at the higher
position is that it alleviates the problem of the valve
outlet becoming submerged when the water level has
risen in the cistern due to a faulty valve, which can lead
to it overflowing. When the outlet is submerged in this
way it is possible that under certain conditions, where
a negative force is acting within the mains supply pipe,
water could by sucked back into the supply, causing water
contamination.

2 With the Portsmouth valve, in order to adjust the water level
in the cistern you must bend the lever arm as necessary.
The modern valve has an adjusting screw to make the
appropriate adjustment to the water level in the cistern.

If you need to replace the float-operated valve for any
reason, it is essential to replace it with the modern diaphragm
type. Repair work on these valves is discussed in Chapter 2.

Head pressure and flow

Finally, before we leave the storage cistern, we will consider the water pressure and volume of water flow that can be expected from the pipe supplying the water.

Pressure is the force of the water. Water pressure can be created by:

* a pump
* a storage cistern positioned high above the water outlets.

Flow is the volume or amount of water passing through a pipe. Water flow is dependent on the pipe size. A 22 mm diameter pipe will clearly allow a greater flow of water than one that is 5 mm in diameter and consequently will fill up an appliance, such as a bath, much more quickly.

The cold water supply feeding your home will be supplied typically via a pump located at the water treatment works. This creates a pressure within your supply pipe of up to around 3 bar (300 kN/m^2). However, when water has been stored in a cistern in your home, possibly located in the loft or roof space, its pressure is no longer influenced by the cold water mains supply but is now dependent on the position of the cistern relative to the taps. The pressure is considerably lower than that in the water mains supply pipe. For example, where a shower takes its water from a storage cistern, there might be only a two-metre head of water, in which case the water pressure will be so low that taking a shower is not practicable. The term 'head' refers to the position of the free water level in the system above the point where it is being drawn off, i.e. in the following example, the water in the cistern is two metres above the shower.

There is a simple calculation that can be completed to find out the pressure created by an elevated cistern. This is: the head of water in metres × 10. So, where the head is only 2 metres the pressure will be:

$$2 \times 10 = 20 \text{ kN/m}^2$$

This is about one-fifth of a bar in pressure ($100 \text{ kN/m}^2 = 1$ bar), and therefore far less than that expected from the mains supply pipe.

From this we can see that a storage cistern should be located as high as possible within a building. Also, the pipe from the storage cistern needs to be a minimum diameter of 22 mm and, where several outlets are to be maintained, it may either need to be increased to 28 mm or a second outlet be taken from the cistern. Failure to observe these simple rules will mean that appliances are very slow to fill.

The toilet flushing cistern

The flushing cisterns used with toilets have undergone several changes over the past 15 years. The water supplied to the cistern is controlled by a float-operated valve. Most of these valves are of a similar design to those used in the cold water storage cistern (Figure 1.6). There are some different designs of valve, but these are beyond the scope of this book.

Prior to 1993, a 9 litre (2 gallon) flush was employed, and it had been like this since the toilet was first designed over 100 years ago. However, in order to try to conserve water this quantity was reduced first to 7.5 litres and then to a maximum of 6 litres, as per current regulations.

In order to discharge this water from the cistern into the toilet pan, a device is used that closes when the required volume has been discharged. Toilet cisterns traditionally worked using a siphonic device (see below), but today there is another design which consists of a valve that is lifted to allow the contents to flow as necessary.

Flushing cistern operated by siphonic action

Siphonic action occurs where water is removed from a container, without mechanical aid, up and over a tube in the form of an upside-down letter J. The long leg joins to the flush pipe, the short leg is open to the water inside the cistern.

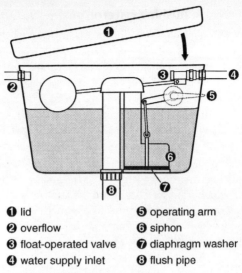

① lid   **⑤** operating arm
② overflow   **⑥** siphon
③ float-operated valve   **⑦** diaphragm washer
④ water supply inlet   **⑧** flush pipe

Figure 1.7 *Flushing cistern operated by siphonic action.*

If the air is removed from the tube a partial vacuum is created. This action, in the case of the flushing cistern, is triggered by the large diaphragm washer being lifted, which discharges a quantity of water over the top of the J-shaped siphon bend. As the water drops down through the flush pipe to the outlet it takes with it some of the air contained within, thus creating a partial vacuum. With the partial vacuum formed, gravity acts upon the surface of the water, pushing down and forcing the water up into the J-shaped siphon tube. As it reaches the top of the upturned bend it simply drops down to the outlet to be discharged into the pan, via the flush pipe. This action continues until the air can get back into the tube to break the vacuum and restore normal pressure. So, water will continue to discharge until the water level has dropped inside the cistern to that of the base of the siphon. The initial action of lifting the diaphragm washer is instigated by the operation of the lever arm located within the side of the cistern.

Valve-type flushing cistern

Several designs of valved flushing cistern have been developed within the past few years; the one shown in Figure 1.8 works by allowing for a dual flush. A dual flush offers:

* a reduced flush for the purpose of removing urine from the toilet pan
* a full 6 litre flush where there are solids to be removed.

There are two buttons housed within the cistern lid, one button with a shorter rod attached to it than the other. When the larger button, with the longer rod, is pressed it lifts the valve sufficiently to engage into a latch and is held up by a small float. Water now flows from the cistern and the latch only releases as the water level drops, taking the float with it. When the smaller button is pressed, the smaller rod does not lift the valve sufficiently to engage with the latch, so the valve is only raised

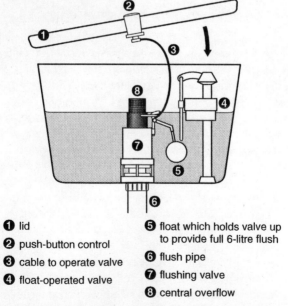

1 lid

2 push-button control

3 cable to operate valve

4 float-operated valve

5 float which holds valve up to provide full 6-litre flush

6 flush pipe

7 flushing valve

8 central overflow

Figure 1.8 *Flushing cistern operated by flushing valve.*

for a short period while the button is held down. A linking cable operates a lever to initially lift the valve from its seating.

Note that a separate overflow pipe is not run from valved flushing cisterns because if the water level rises, due to the water inlet failing to close, it would overflow down through a central core within the valve from the cistern and into the toilet pan.

Hard and soft water

Water is generally classified as being either hard or soft. This classification relates to the impurities that the water contains and is indicated as a measure of the number of hydrogen ions (acidic) or hydroxyl ions (alkaline) present in a sample of water. This is known as the 'potential of hydrogen' value (pH value):

pH value of water

1 2 3 4 5 6 7 8 9 10 11 12 13 14

soft (acidic) ⟶ (alkaline) hard

neutral

Hard water contains calcium carbonate and/or calcium and magnesium sulphate, which basically means limestone in one form or another, whereas soft water does not. This limestone has dissolved in the water because it is a natural solvent. The hardness of water can be further classified as:

* permanently hard (contains dissolved rock such as calcium or magnesium sulphate)
* temporarily hard (contains dissolved rock such as calcium carbonate).

The limestone in permanently hard water cannot be removed without water-softening treatment. Temporary hardness, however, occurs where the rainwater has fallen onto calcium carbonate. This is a different form of limestone and will only dissolve in the water if it contains carbon dioxide, which the water acquired as it fell as rain. Boiling the water can remove the temporary hardness

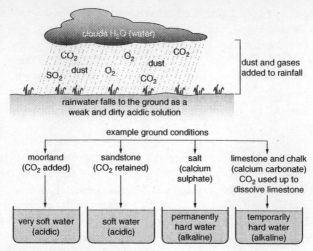

Figure 1.9 *Formation of soft and hard water.*

because the carbon dioxide escapes from the water, but boiling will have no effect on permanently hard water.

Soft water, on the other hand, does not contain any dissolved limestone. It is more acidic or aggressive as a solvent, and will soon destroy metals, in particular lead when used within a plumbing system. Soft water feels different from hard water and is more pleasant to wash in; it is also much easier to obtain a lather when using soap in soft water, and it takes longer to rinse the soap away. Hard water is also distinguishable by the scum that forms on the surface of the water and around sanitary appliances, and by the limescale that forms around taps and in the toilet bowl.

Limescale

Limescale can be found at the outlet point of both hot and cold taps in hard-water areas. What is it? In a nutshell, it is caused by temporarily hard water. However, a more in-depth explanation is appropriate here.

When it rains, the water falling from the sky is enriched with carbon dioxide (CO_2), trapping it within its molecular

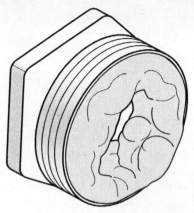

Figure 1.10 *Limescale build-up in a pipe reducing the size.*

structure. This water falls to earth and percolates through the ground on its way to the rivers and reservoirs. If it flows through limestone during this journey, the CO_2 in the water causes the limestone to dissolve and, as a result, the limestone is carried within the water. When the CO_2 escapes from the water, such as by rapid shaking movements or by heating the water above 60°C, the limestone will not remain dissolved, as it was the CO_2 that maintained this condition. Consequently, the limestone comes out of the water and collects within the system as solid limestone (limescale). It is found around tap heads because the water collects here as it leaves the spout, and as the water evaporates, the solid limestone is left behind.

It is important to note that limescale can also collect, unseen, inside your pipework, accumulating around heating elements and heat-exchanger coils, causing long-term damage and affecting heat-up times. It can drastically reduce the rate of water flow through any pipes in which it collects. To prevent this, it is essential to store the water at a temperature no higher than 60°C, as above this temperature the CO_2 can more easily escape from the water. Limescale can also be prevented by using water softeners and conditioners.

A water softener

This is a device that is designed to remove all of the calcium and magnesium ions from the water. Basically, the water is passed through a bed of special chemical called zeolite, or through very small plastic beads covered with sodium ions, and as a result, the calcium and magnesium are given up. However, the zeolite bed eventually becomes exhausted to the point where it stops softening the water. It is then time to regenerate the bed material with sodium ions. This is achieved by passing a salt solution (sodium chloride) through the softener to displace all of the calcium and magnesium and recharge it with sodium ions. The regeneration process flushes out all of the unwanted products into a drain. The process of regeneration is completed automatically, timed to take place during the early hours of the morning; during this period no softening takes place and hard water will be supplied if a tap is turned on. A water softener is the only device that removes the calcium and magnesium from the water.

A water conditioner

This is not a water softener but a device that reconditions the small dissolved particles of limestone, referred to as calcium salts, held in suspension in the water so that they do not readily stick together to form noticeable limescale. If you viewed untreated hard water under a microscope, the calcium salts would appear star-shaped, with jagged edges. It is in this form that they stick together. The water conditioner aims to take off these jagged edges so that they cannot easily bind together and instead they simply flow through the system. There are two basic types of water conditioner. First there are chemical water conditioners, which use crystals that dissolve in water and bind to the star-shaped salts, sticking in the crevices and jagged edges and having the effect of rounding off the sharp points.

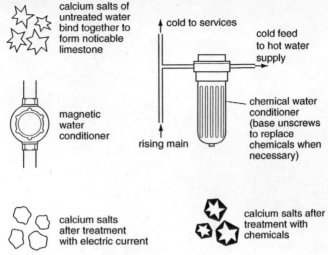

calcium salts of untreated water bind together to form noticable limestone

cold to services

cold feed to hot water supply

magnetic water conditioner

chemical water conditioner (base unscrews to replace chemicals when necessary)

rising main

calcium salts after treatment with electric current

calcium salts after treatment with chemicals

Figure 1.11 *Water conditioners.*

The other type of water conditioner passes a small electric current of a few milliamps across the flow of water. This current alters the shape of the calcium salts, changing them to a smoother and more rounded shape. This current is often produced by a magnet, although other methods can be used.

The above-ground drainage system

The first thing water does as it goes down the plug hole is to pass around a range of bends that form a small trap of water. You can see this trap by looking into your toilet pan or beneath the kitchen sink. Why is the trap there? It is not there to catch your wedding ring should it come off your finger, although this function can prove useful in such a circumstance. Its purpose is to provide a pocket of water between the outside air and the foul air within the drain and sewer. This air would most certainly be foul-smelling and may also contain methane gas, which could prove hazardous. Another purpose of the trap is to prevent any

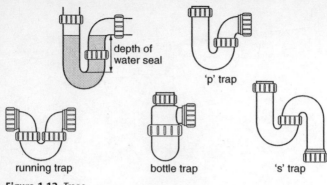

depth of
water seal

'p' trap

running trap   bottle trap   's' trap

Figure 1.12 *Traps.*

vermin that may be in the drain from entering the building. This trap is the start of the house waste water system.

Gravity causes the water to flow from the trap along pipes that run down to adjoin the vertical discharge stack, referred to as the soil and vent pipe, and from here all the various waste pipes converge to take the fluid to the drainage system below ground. Obviously, the pipe must always be laid to fall in the direction of the water flow and the pipe must never, under any circumstances, be run uphill as water simply will not drain from the pipe.

The system illustrated in Figure 1.13 is generally referred to as the single-stack system, although it is given the fancy title of 'primary ventilated stack system'. This has been installed in homes now for more than 60 years.

There are many houses around that are much more than 60 years old, so there are systems in existence, such as that shown in Figure 1.14, that have a separate waste water discharge stack and foul-water stack. It was not until the pipes reached the ground-level drain that they were joined together. When major refurbishment to these antiquated systems is undertaken the plumber will update the system and install a single-stack system.

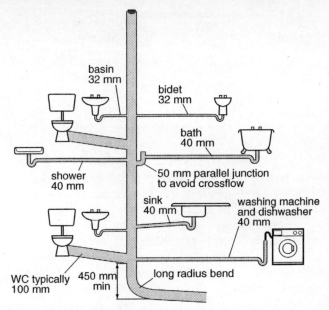

Figure 1.13 *Typical primary ventilated stack (single-stack) system.*

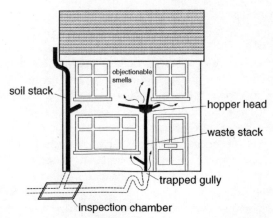

Figure 1.14 *The older system of separate waste stack and soil stack.*

Table 1.1 *Maximum lengths for discharge pipes.*

Pipe size	Maximum length
32 mm	1.7 m
40 mm	3.0 m
50 mm	4.0 m
100 mm	6.0 m

Plastic pipework is used for modern systems. This will either be of a type that can simply be pushed together, or the joints can be made using special solvent weld cement, which bonds the pipe to the fitting. The pipe diameters are shown in Figure 1.13. The lengths of the pipes from the mains stack should be limited and this distance should not exceed the distances listed in Table 1.1, otherwise you may experience problems with self-siphonage, explained below. It should also be noted that the flow of water passing horizontally to the vertical stack has been run to a minimal fall usually not exceeding a drop of between 18 mm and 90 mm per metre run of pipe. Exceeding this gradient could also create self-siphonage problems and can increase the problems of leaving any solid contents behind as the water rushes rapidly down the pipe.

Water siphonage from the trap

Water being siphoned from a trap is recognized by a gurgling sound coming from the appliance as air tries to enter the waste system in order to maintain the equilibrium of air pressure from inside the pipe to that of the surrounding atmosphere. Two types of siphonage can be encountered (Figure 1.15):

* Self-siphonage – caused as the water flows through the pipe, forming a plug of water, causing a vacuum to be formed, sucking with it the water from the trap.
* Induced siphonage – occurs when no water has been discharged and is simply the result of a design error caused by the installer joining two waste pipes together so that

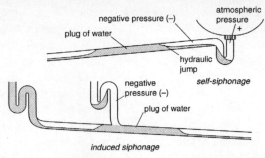

Figure 1.15 *Water siphonage from the trap.*

as the water of one appliance flows past the branch connection of the other, the air is drawn from the pipe.

Where continued problems are encountered with siphonage it is possible to fit:

* a resealing trap – which uses the concept of incorporating a non-return valve
* a special trapless (self-sealing) waste valve – sold under the manufacturer's trade name of HepvO, this contains a special synthetic seal instead of the traditional water seal, which closes in the absence of water to seal off the pipe.

Air admittance valves

Another device sometimes used to overcome problems with siphonage is an air admittance valve. This is basically like a big

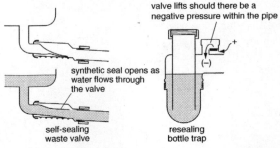

Figure 1.16 *Alternative trap designs.*

non-return valve that allows air to go into the drainage system but prevents air (potentially foul-smelling) from coming out. So where a negative pressure exists inside the drainage system, this valve opens in preference to the water being sucked from the trap.

Air admittance valves can be purchased in a whole range of sizes, and sometimes the main discharge stack itself is terminated with an air admittance valve, possibly found within the roof space. This fitting is generally used where there are two soil stacks within the same building or where there are several buildings in close proximity. It overcomes the need to run the highest point of the discharge stack out through the roof, avoiding additional work to the roof tiles and ensuring that rainwater cannot enter the building.

An air admittance valve must be fitted above the spill-over level of the appliance (the highest possible water level of the nearest adjacent appliance), otherwise if there is a blockage in the pipe, this fitting will be subject to a backup of water and the valve is unlikely to remain watertight.

Where these valves are in exposed locations, such as in the loft, they do need to be insulated to ensure that they do not freeze up, as there is often a considerable amount of condensation within the pipe.

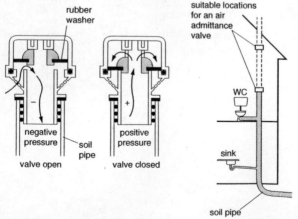

Figure 1.17 *Air admittance valve.*

Access points

All good systems of drainage should have a means of access for internal inspection of the pipe, which is particularly useful when there is a blockage. Sometimes a large access point is positioned to the end of a small vertical section of 100 mm diameter discharge pipe, used as an alternative to the air admittance valve for an additional ground-floor toilet within the property. This method is acceptable provided that the pipe lengths are not excessive and, in all cases, no further than 6 metres from a ventilated drain, otherwise additional pressure fluctuation problems will be created within this section of pipe.

As with the air admittance valve, this access point must be installed above the spill-over level of the appliance, otherwise if there is a blockage to deal with, when it is opened, the foul water will discharge all over the floor.

Remember that if at any time you need to open an access point, you must consider what might lie behind! If water is there at a time of blockage – which may be the reason for opening this access point in the first place – it is likely to flow uncontrollably, at surprisingly high pressure, on to you and the floor where you are standing.

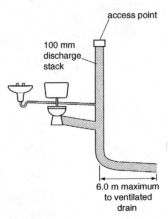

Figure 1.18 *Access point.*

Pumped sanitation and drainage systems

For many years now, there has been the opportunity to locate a drainage point for the purpose of removing water from basins, shower units and even from WC pan connections, from more or less anywhere within a typical house. These systems use what is called a macerator pump. This is basically a small holding tank with the additional facility to macerate (chop up) any solid matter within, and which, when full of water, operates a pump to lift the water contents up or along through a small pipe (typically no bigger than 22 mm) to discharge into a drainage stack.

The manufacturer's data sheets should be sought for the various designs but typically the water could be elevated vertically by 4 metres, and horizontally the water could be discharged up to 50 metres. One final point on the installation of these units is that it is a requirement that the property also has a conventional gravity system of drainage from a WC, otherwise, if the power to the building is off due to a power cut, you would be without a toilet.

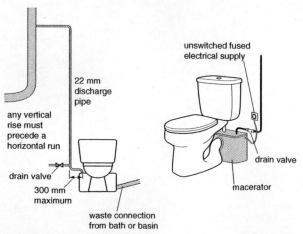

Figure 1.19 *Pumped sanitation system.*

The water closet (WC)

The term water closet technically refers to the room in which a toilet pan is found. But when talking of the WC, one is generally referring to the complete package of toilet cistern and attached pan.

The WC suite has undergone several design changes over the past few years. Today, Water Regulations limit the volume of water flushed down a newly installed toilet pan to a maximum of 6 litres, yet not many years ago this volume was 9 litres. Most toilets installed these days are of the wash-down type, which basically means that they rely on the discharging water flow to remove the contents from the pan.

Occasionally, siphonic WC pans will be found. These were installed quite extensively during the 1970s but are becoming rare these days as people update their homes. The siphonic pan, however, had one advantage over the wash-down pan in that it had the additional siphonic action to assist the removal of the pan's contents. It basically worked by lowering the air pressure from the pocket of air trapped between the two traps. This was achieved by allowing the flushing water to pass over a pressure-reducing fitting which created a negative pressure

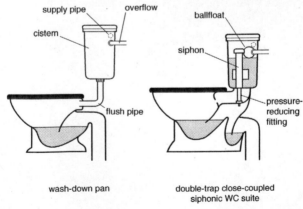

wash-down pan

double-trap close-coupled
siphonic WC suite

Figure 1.20 *WC installations.*

and sucked out the air between the two traps of water. With the partial vacuum created, the water and its contents in the upper bowl of the pan were sucked out by siphonic action. The cost of manufacture is possibly the reason for their disappearance.

The below-ground drainage system

Once the water has reached ground level it is conveyed to the house drain, which removes it from the property to meet up with the public sewer, or the water may be collected in a septic tank or cesspit.

Septic tank

This is a private sewage disposal system used in some rural areas. Basically, all the foul and waste water is collected within a large double-compartment chamber, traditionally made of brickwork or concrete, although nowadays these are generally made from plastic. From here the water overflows through an irrigation trench to slowly filter into the ground away from the property.

These systems rely on a scum forming on top of the liquid and in so doing allow anaerobic bacteria to decompose most of the solids. Because not all the solids are broken down, it is necessary to have the vessel emptied annually to remove the

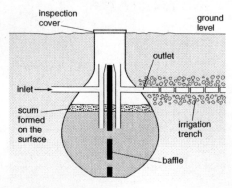

Figure 1.21 *Septic tank.*

accumulation of the excess sludge that will not decompose. Failure to do so may lead to a blockage in the system.

Cesspool (cesspit)

This is simply a watertight container that is used to collect and store waste and foul water from the property. Cesspools are used where no mains drainage has been connected to the property and there is insufficient provision for a septic tank. The tank will need to be emptied, ideally before it is full, by a contractor, for proper disposal.

Surface water

In addition to the water that flows into the drains from the various sanitary appliances in the home, water is also collected from the gutters, rainwater pipes and large paved areas – this is generally referred to as surface water. If the drain is serviced by a septic tank or cesspool it will require an additional separately run drain for the purpose of collecting the surface water, because if this water is allowed to flow into these holding tanks it will cause them to fill too rapidly. In these cases, the surface water might be collected and run into a drainage ditch, river or soakaway.

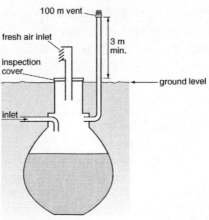

Figure 1.22 *Cesspool.*

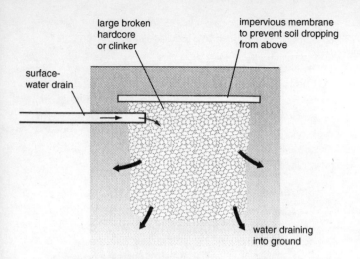

large broken
hardcore
or clinker

impervious membrane
to prevent soil dropping
from above

surface-
water drain

water draining
into ground

Figure 1.23 *The soakaway.*

The soakaway is simply a large hole filled with rubble, into which the drainpipe runs. The water collects here and gradually drains into the surrounding ground.

Connections to public drainage systems

If the foul-water drain is connected to a public sewer, the surface water may be collected within the same pipe and they run off from the property together. This is referred to as a combined system of drainage. Whether or not a combined system of drainage is used will depend very much upon the local authority, which treats all the water. As a consequence, some areas have separate systems of drainage, in which the surface water is run into its own specific pipe.

When making any new connection to a drainage system it is essential to confirm the type of drainage system you have. Failure to do so could result in contamination of the local water course if you inadvertently discharge foul water into a surface-water drain.

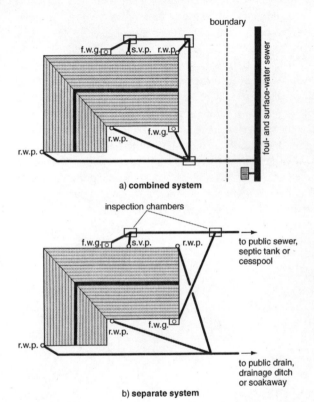

r.w.p. – rain water pipe
s.v.p. – soil and vent pipe
f.w.g. – foul-water gully

a) **combined system**

b) **separate system**

Figure 1.24 *Connections to public drainage systems.*

In addition to the systems identified in Figure 1.24, there is a slight variation that can be found where an occasional surface-water connection is a long way from the surface-water drain, or there is some difficulty in getting past the drainage pipe of a foul-water drain. In such a situation it is possible that this one-off connection can be discharged into the foul-water drain.

If this is done, the system is referred to as partially separate; however, it must be understood that no cross-connection can be made the other way around, i.e. the foul water must *never* be allowed to connect to the surface-water drain.

Where a separate system of drainage is employed, the connections of the pipes to the surface-water drain do not have to include a trap. However, all connections made to the foul-water drain, be it surface water or waste water, must be trapped. If you look carefully at Figure 1.24 you will see foul-water gullies (f.w.g.). These are traps at ground level, 100 mm in diameter, i.e. the same size as the house drain, into which smaller pipes have been run. Prior to the 1970s these traps were left open, with a small brick course around the opening and a grate above the water level; nowadays the pipes entering these gullies are discharged below the ground surface into a side inlet pipe, and an access cover is secured at ground level.

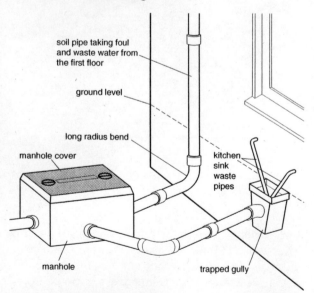

soil pipe taking foul
and waste water from
the first floor

ground level

long radius bend

manhole cover

kitchen
sink
waste
pipes

manhole

trapped gully

Figure 1.25 *Connection of the above-ground drainage into the below-ground drainage system.*

The soil and vent pipe connected to the drain is not trapped. However, it should be noted that all appliances connected to this pipe are themselves trapped. This pipe allows the free passage of air into and out of the drain, thereby maintaining equal air pressures within the drain and outside it. Air flowing through the drain also assists in drying out any solid matter left behind during flushing; as it dries it shrinks and is more easily flushed away during the next discharge of water.

Gutters and rainwater pipes

This is the last part of the plumbing system – the 'outside plumbing' – and the only part that if it was to leak would make little difference. The guttering consists of a simple channel located at the base of a roof to catch the run-off of water. From here the water runs to the outlet and falls down the rainwater pipe to the surface drainage system below. Forty years ago metal was used for the installation of this last part of the plumbing system but, like so many things today, plastic has long since replaced these older traditional materials.

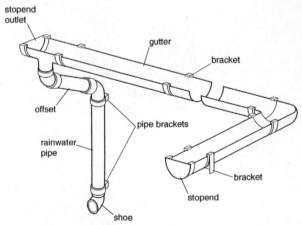

Figure 1.26 *Gutters and rainwater pipes.*

2

hot water in your home

Within this chapter the various types of hot water supply that will be encountered within a home are identified. There are many different designs; some take their water directly from the cold water mains supply, whereas others are supplied with water via a storage cistern. Some systems heat only the amount of water that is required, whereas others heat vast amounts of water and store it within a hot storage cylinder for use as and when required. The design of your hot water system will mostly depend upon the age of your building. The most common systems include:

* a gas or electric single-point water heater found above the sink or basin
* a gas multipoint water heater that serves all the hot water outlets
* a boiler used to store hot water within a cylinder; this system may also serve the central heating
* a combination boiler to provide both central heating and hot water instantaneously
* a thermal storage system (by far the least common).

These systems are classified as being either:

* storage (vented or unvented)
* instantaneous (combination boiler, multipoint, single-point or thermal storage).

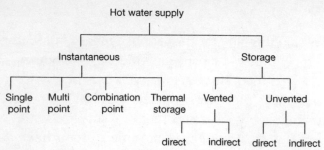

Figure 2.1 *Types of domestic hot water systems.*

Hot water storage systems

Domestic hot water is stored in an enclosed vessel, which is most likely to be a cylinder, suitably insulated to keep the heated water warm. This vessel is found typically in an airing cupboard. The water is heated either directly or indirectly.

The installation of modern domestic hot water systems is controlled by legislation, which is particularly rigorous with regard to energy efficiency. If you want a new gas or oil boiler to use with a hot water cylinder, you cannot just install any old appliance. It must conform to the standards laid down within the Building Regulations, which are administered by the local authority. Consequently, when a boiler or cylinder is replaced, the local authority may wish to be notified in order to ensure that it is in compliance with current standards.

Storage cylinders have developed and become more efficient over the years. Older cylinders:

* required a cylinder jacket to be tied around them in order to keep as much heat as possible from being lost to the surrounding space. They were usually installed in a cupboard, which stayed warm and dry and thus provided an ideal storage area for airing clothes. However, in this modern age of energy efficiency they have been identified as using fuel inefficiently

* had 1½–2 turns in the internal pipe coil that made up the heat exchanger. This led to a very slow heat transference rate and increased the time taken to heat the water in the cylinder as it passed from the primary heating circuit.

Modern cylinders:
* are foam lagged at the manufacturing stage
* have at least 5–6 turns in the heat exchanger, increasing heat transference times.

It is also possible to purchase high-performance cylinders that have a bank of many coils passing through the cylinder, allowing for even faster heat-up times.

If you have an old style of boiler, it may be worth considering replacing it with a new one next time it needs any repair or maintenance work. This will reduce the time it takes to warm up the water and will in turn save money and provide better fuel efficiency.

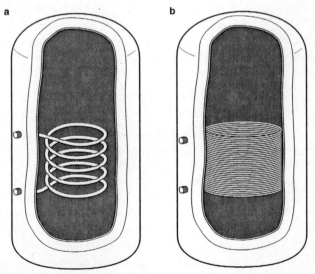

Figure 2.2 *(a) Normal cylinder heat exchanger coil; (b) high-performance cylinder heat exchanger coil.*

Water temperature

The temperature of the hot water is set by the installer and should be adjusted to meet the needs of the end user. The temperature within a stored hot water cylinder should be adjusted to no higher than 60°C at the top of the cylinder. If it is set higher than this, the water may scald the user, and also limescale deposits may form in hard-water areas. Equally, the water should not be stored at a temperature much below this as the growth of legionella bacteria may occur.

Legionella

Legionella is rarely a problem in domestic homes. The bacteria are killed off above 60°C and will not survive very long at this temperature. However, they can survive within the temperature range of 20–45°C. They can be dangerous to humans and are transmitted when water is in a misty or vapour form, so areas around boosted shower outlet sprays could be vulnerable if the water is maintained at too low a temperature. The best alternative to use where cooler water temperatures are required is to store the water at 60°C and then use a blending/mixing valve, which mixes the hot water with a quantity of cold water to reduce the temperature to the desired level.

Direct systems of hot water supply

As the name suggests, direct systems include those in which the water is heated directly, such as by an immersion heater or by a boiler found some distance from the hot water storage cylinder. The heated water is transferred to the cylinder by gravity circulation (Figure 2.3) via two pipes referred to as the primary flow and primary return. Where the water is heated in a boiler it is invariably quite hot, and limescale build-up will occur inside the primary pipework in hard-water areas. Most direct systems are now quite antiquated and only the oldest of houses will still have such a system. The immersion heater, however, is still quite commonplace and makes an ideal backup

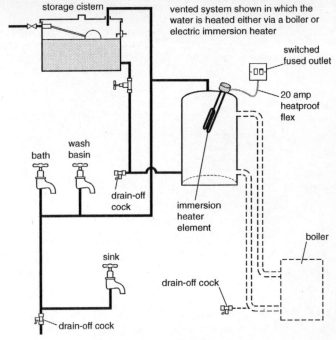

storage cistern

vented system shown in which the
water is heated either via a boiler or
electric immersion heater

switched
fused outlet

20 amp
heatproof
flex

bath

wash
basin

drain-off
cock

immersion
heater
element

boiler

sink

drain-off cock

drain-off cock

Figure 2.3 *A direct system of hot water supply.*

when incorporated within the cylinder of an indirect system of
hot water supply.

The immersion heater

This is effectively a large heating element like those found
inside a kettle. When the immersion heater is switched on, the
element heats up and remains on until the thermostat senses
that the water temperature has reached its desired level or until
the power is switched off. As mentioned earlier, the water should
be stored no higher than 60°C; this level is set by making an
adjustment to the dial on the head of the thermostat. Where the
immersion heater is fitted within an unvented hot water cylinder
it will also require a high-limit cut-out thermostat set to operate

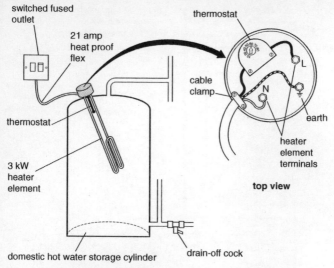

Figure 2.4 *The immersion heater.*

(cut out) at 90°C. All new and replacement immersion heaters will include, as standard, this independent non-self-resetting over-temperature safety cut-out device to prevent the water in the cylinder from overheating.

Gravity circulation

The hot water from the boiler in Figure 2.5 is transferred to the cylinder by natural gravity circulation, i.e. hot water rises up the primary flow and is displaced by the column of descending cooler water within the primary return. This system is very common and will be found in a large number of properties. However, these systems are slow and the water in the cylinder can take anything up to two hours to heat up. Modern systems use a circulating pump to push this water around the circuit rapidly, allowing heat-up times of around 30 minutes, or sometimes even less (see Chapter 3 for examples of fully pumped central heating systems).

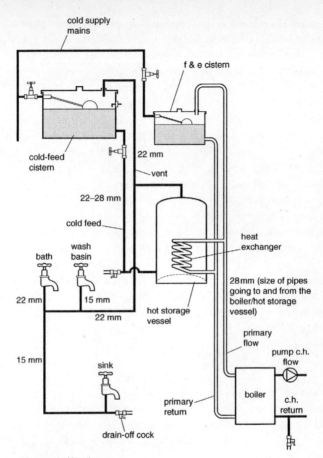

Figure 2.5 *An indirect system of hot water supply.*

c.h. = central heating
f & e = feed and expansion

Indirect systems of hot water supply

If you have a hot water cylinder in your home, there is a good chance that it is part of an indirect system. With this type of system there are no problems with hard water scaling up the pipes, and central heating water can also be taken from the water heated within the boiler.

Indirect systems of hot water supply have a heat exchanger coil located inside the hot water cylinder. This is, in effect, a pipe run in a series of loops inside the cylinder of water. Hot water from a boiler is passed through this pipe and the hot water flowing through the pipe coil in turn heats up the water in the cylinder. Thus the water is heated directly within the boiler, as in the direct system – referred to as the primary hot water – and indirectly via the pipe coil within the cylinder – referred to as the secondary hot water or domestic hot water (dhw).

Indirect systems may be either vented or unvented. Vented systems are those in which the cold water is taken from a cold water feed cistern, usually found in the roof space; unvented systems are fed with cold water directly from the cold supply mains pipe. As can be seen in the vented system in Figure 2.5, there are two separate cisterns within the roof space or loft. One is the cold water feed cistern, designed to supply water to the cylinder, and the other is a feed and expansion (f & e) cistern. An unvented system can be seen in Figure 2.6.

Vented systems

Feed and expansion cistern (f & e cistern)

The f & e cistern ensures that the water contained in the boiler and heating system, where applicable, does not mix with the water used for the domestic hot water. There are two specific reasons for this separation:

* to combat the problem of limescale build-up
* to reduce the amount of atmospheric corrosion.

In domestic hot water pipework, water is constantly passing through the system and this constant flow of oxygenated water contains a quantity of dissolved limescale. However, as illustrated in Figure 2.5, the water that enters the boiler and heating system via the f & e cistern – which is heated to far in excess of 60°C – is never emptied unless it is drained out for maintenance purposes.

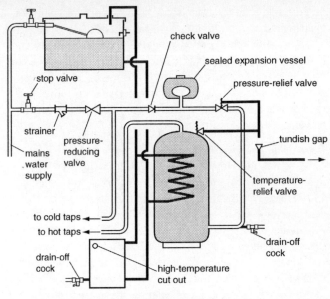

a) **system using a sealed expansion vessel**
(showing the water heated within a boiler)

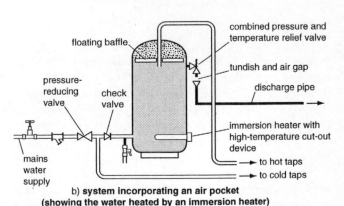

b) **system incorporating an air pocket**
(showing the water heated by an immersion heater)

Figure 2.6 *Unvented systems of domestic hot water supply.*

So, limescale build-up is eliminated because once the water has been heated, no more limescale will be generated.

Also, after a short period of heating the water and moving it around the system with a circulating pump, all of the air will have been expelled from the system. It is this air, in particular the oxygen in it, that causes the corrosion of iron, from which the boiler and radiators are invariably made, so losing this air prevents them from rusting.

The water level within the f & e cistern

The water level within the f & e cistern is adjusted low down inside this cistern, just above the outlet. As the water within the system heats up it expands, rises back up the cold-feed pipe and is taken up into this cistern. If, during the installation of these cisterns, the water level is adjusted too high, the water, when heated, will expand and rise to a point where it will drip from the overflow pipe. Upon cooling, more water will re-enter the cistern via the float-operated valve and the process of overflowing will be repeated continually. This will result in fresh oxygenated and calcium-laden (limescale-forming) water continually being added to the system.

The open vent

You may be wondering why a pipe with an open end terminates above the water level within the cistern. Why is the vent pipe needed? First, it allows air in and out of the system during filling or draining down. You will notice that in Figure 2.5 the water enters low down in the cylinder, near the bottom, and the hot water is drawn off from the top. If there were no vent pipe there would be a very large air pocket above the water which would prevent the water from filling the system. Also, when draining out the water from the system the vent pipe allows air to enter, which makes it easier to remove the water.

The second purpose of the vent pipe is as a safety measure, ensuring that the system always remains at a pressure compatible with that of the atmosphere and allowing any pressure generated within the system to escape. A build-up of pressure could result from the cold feed to the system being blocked, as might happen if it freezes in winter or if debris accumulates inside the base of the storage vessel.

Hot water distribution

If you look again at the example of stored hot water supply (Figure 2.5), you will notice that the hot water is drawn off from the top of the cylinder. The reason for designing the pipework in this way and taking the water from the top of the cylinder is that this is where the hottest water is found, because hot water naturally rises to the highest point. The cold water flows in at the base of the cylinder and pushes the hot water out when a tap is opened. If the cold water were supplied at the top of the cylinder it would mix with the hot water and cool it down.

Some cylinders are designed with the cold pipe connected at the top, which would appear to contradict this argument but, in fact, if you could see inside the cylinder you would notice that the pipe extends inside the vessel, right down to the base – an example of a 'dipped cold feed', as it is called, can be seen in Figure 2.12.

Water expansion

When water is heated it expands by approximately 4 per cent from cold to 100°C. (Above 100°C, at atmospheric pressure, it changes to steam and its volume immediately expands 1600 times.) For safety reasons the expansion of the water must be allowed for in the design of the storage cistern.

If you have an open-vented system, it will be under the influence of atmospheric pressure and as the water slowly heats up it will expand and be pushed back up through the cold-feed pipe into the cold-feed cistern that supplies the system. As mentioned above, if the cold feed becomes blocked, the expanding water will be forced to travel up the open vent pipe and discharge into the cistern, thereby preventing a pressure build-up within the system.

Imagine the possible danger if both the cold feed and the vent pipe became frozen up and blocked. If the water were to heat up and expand, this expansion could not be accommodated and, as a result, the cylinder might split at the seams or even explode, hence the need to ensure that pipework is suitably insulated.

Unvented systems of hot water supply

Many homes built today incorporate this design of hot water supply. It has the advantage of:

* having a stored supply of potable hot water
* maintaining a good flow rate to the various outlet points
* being at water supply mains pressure
* freeing up the roof space to assist in the design of modern roof structures.

This type of system has only been permitted since 1985 and, as a result, is generally only found in newer developments or houses that have been refurbished. It is essential to note that the minimum size of the supply pipe to these systems is 22 mm – if it is any smaller you will not get the flow rate expected as compared to that of the vented system with its increased pipe sizes. New homes are constructed with this larger mains supply pipe, thereby generally posing no problems; existing properties, however, may only have a 15 mm inlet cold water mains supply and this will be inadequate to serve all of the hot and cold outlets within the property.

The installation of these systems falls within the requirements of the Building Regulations, as administered by the local authority, so the installation and maintenance of these systems must be undertaken only by approved operatives. This means that the installer will have taken and passed an assessment course aimed specifically at the design and safety of these systems.

Looking at the two systems shown in Figure 2.6 you will see that there are several controls in addition to those found on the more traditional systems (Figures 2.3 and Figure 2.5). Two systems have been illustrated because one design uses a sealed expansion vessel to take up the expanding water, whereas the other uses an air pocket, located inside the cylinder with a floating baffle.

The following notes provide a brief outline of the controls found on an unvented system, purely for interest and identification but, as stated above, remember that it is a requirement that these systems are only installed and serviced by qualified personnel. Should you have such a system and require work to be completed on these controls, remember to ask the operatives to show you an approval certificate or card, otherwise your house insurance may not be valid should something go wrong, as these systems can explode if not looked after properly.

Components of the unvented system

Strainer

This is designed to ensure that no grit or dirt within the pipeline can travel along the pipe and cause the ineffective operation of a control installed further downstream.

Pressure-reducing valve

This is a special control that prevents excess mains pressures from entering the hot water storage vessel. The hot water storage vessels themselves are quite robust but will not withstand the highest water pressures sometimes experienced within the mains supply. This control usually restricts the pressure to a maximum of 3 bar. In order to ensure equal pressures in both hot and cold supplies, such as where mixer taps are incorporated, the cold water is sometimes branched off after this control valve, as seen in Figure 2.6. Alternatively, a second pressure-reducing valve will be required on the cold-supply pipework.

Check valve

This valve is basically a non-return valve that has been incorporated to prevent the heated water expanding back along the pipework. It is a Water Regulations requirement that no water is allowed to flow in a direction opposite to that intended.

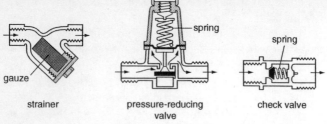

Figure 2.7 *Components of the unvented system.*

Sealed expansion vessel

This unit is designed to take up the expanding water within a large rubber bag contained within an airtight vessel. As water is heated it expands and flows into the bag. This causes the air surrounding the bag to become pressurized; when the water cools, the air pressure forces the water back out into the system. Note that some systems do not use the sealed expansion vessel as identified here but take up the expanding water within an air pocket located inside the top of the cylinder.

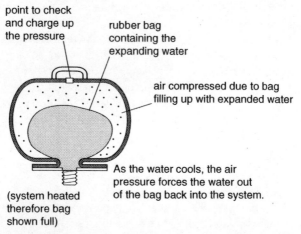

Figure 2.8 *Sealed expansion vessel.*

High-temperature cut-out thermostat

This is basically a second thermostat in addition to the normal thermostat. This control will turn off the supply when the temperature within the system rises to 90°C. Should this control be activated you will need to manually reset the device.

Pressure-relief valve

This is a special control valve designed to open, allowing water to discharge from the system into a drain, should the pressure rise to such a point that damage to the storage vessel might result.

Temperature-relief valve

This is another special control valve that is designed to open should the high-temperature cut-out device fail to work. It allows the water to discharge from the system safely into a drain should the temperature rise to around 95°C, at which point it would become dangerous. With the high pressures that might be generated within the system due to heating water, the boiling point is increased and if the temperature were to get any hotter than this, uncontrollable steam could discharge from this control rather than controllable water.

Sometimes the pressure-relief and temperature-relief valves are incorporated within the same control valve and in both

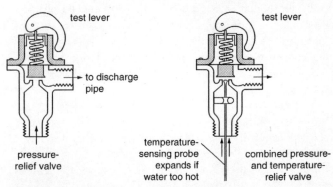

Figure 2.9 *Pressure- and temperature-relief valves.*

cases any water discharging from them is conveyed to the drain via an air gap and funnelled tundish. The air gap is maintained to ensure that the drain pipework cannot make contact with the potable hot water supply pipework.

Instantaneous systems of domestic hot water supply

The storage systems discussed above work well, and a good flow rate of water from the taps can be expected from a correctly sized system. However, in the case of unvented systems for homes with many occupants or older properties with a small inlet supply pipe – which might be just 15 mm in diameter – an instantaneous system may be the only choice where a connection to the cold mains supply pipe is made and has very much been the traditional system of domestic hot water supply.

Older properties that do not have heating systems often have an instantaneous system of hot water supply. They may have a centrally installed multipoint water heater or several single-point water heaters found at the appliances where the water is required. These heaters may be electrically operated or could also be gas fuelled.

Many homes have upgraded from the multipoint system by installing a combination boiler (often called a combi boiler for short) to supply hot water and central heating. These units heat water as it is required, rather than storing it at high temperatures, and also provide hot water that can be used for heating purposes.

The biggest drawback with the instantaneous water heater is the fact that the water can only be heated at a limited rate and, as a result, the flow rate from the outlet tap is invariably slower than that expected from a storage system. The layout of the pipework to the various appliances is, however, the same as shown in Figure 2.10 below.

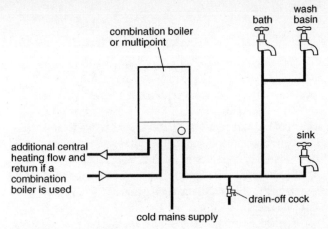

Figure 2.10 *Centralized system of instantaneous domestic hot water using a combination boiler or multipoint.*

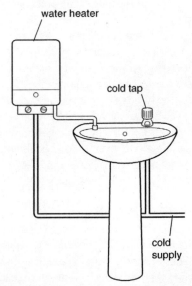

Figure 2.11 *Localized single-point instantaneous hot water heater at the point of use.*

Thermal storage systems of hot water supply

These systems were introduced around 1985 as an alternative to the unvented storage system, having a supply of hot water at mains pressure, without all the necessary safety controls required for unvented systems. They are, in effect, a system of instantaneous hot water supply, taking their water directly from the mains supply. The difference between this and the unvented system is that this does not store hot water used for domestic draw-off purposes. Unvented systems are classified as such due to the fact that they contain a stored volume of water in excess of 15 litres.

If you look closely at Figure 2.12 you will see that the storage cylinder is full of hot water, but it is not used for a domestic supply to the taps, as with all the storage systems previously identified; it is only used to supply the heating circuit and thereby used to warm the radiators.

In the hot water container there is a pipe coil heat exchanger with many loops. If a hot water tap is turned on, the water will flow directly from the water mains through this coil, which causes the water to heat up rapidly, taking its heat from the cylinder full of hot water. It then passes through a blending/mixing valve which allows a percentage of cold water to mix with it if necessary, thereby cooling it to the desired temperature, as it may have become too hot when passing through the cylinder heat exchanger. This system is by far the least common, but it is found in some homes.

Hot distribution pipework

Whether the centralized domestic hot water system is of the storage or instantaneous type, the water must flow around the building in pipes of appropriate size, reducing down in size to the smaller pipes that serve the various outlet points. A pipe of minimum diameter 22 mm needs to be used to supply a bath.

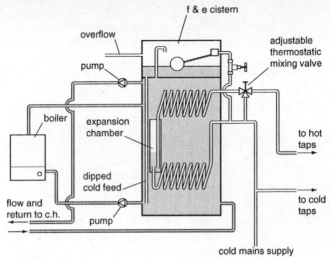

Figure 2.12 *Thermal storage system of domestic hot water supply.*

As with the cold water pipework, a drain-off cock is located at the lowest point of the hot water pipework in order to facilitate draining down if necessary.

Choice of domestic hot water supply

What is best, a combination boiler or a regular boiler with a storage cylinder? This is a question that you will ask yourself when considering a new hot water supply. Each system has its own merits, and when designing a system you should weigh up the pros and cons in order to choose what is best for you. Some of the merits and pitfalls of each system are discussed below.

Combination (combi) boiler

A combination boiler heats up water for domestic use, providing hot water for the taps and for the central heating system. The installation of combination boilers currently makes up 60 per cent of market sales and therefore deserves the first consideration. However, it will not always be the best choice. The clear advantages include:

* easily installed and is the cheaper option
* only heats the water as and when it is required
* does not require a storage cylinder or cistern in the roof space
* water fresh from the mains is used for the hot supply to the taps
* the water will be at a good pressure for showers
* provides water for central heating.

There are lots of good points here, but this system also has disadvantages that are often overlooked when considering the installation of a new system of hot water supply. These include:

* a poor flow rate from the taps where the pipe size to the house is inadequate
* no boiler operation for central heating purposes when it is being used to heat the water for domestic hot water
* there will be no backup supply of hot water if either the power or the water supply is turned off.

Let's look more closely at these disadvantages. First, consider that the pipe entering the property is only 15 mm in diameter – you just might be expecting too much from the pipe. The modern home often has dishwashers, washing machines, outside taps, numerous toilets and bathrooms. You cannot possibly expect this one pipe to feed all of these outlet points at once. It is unlikely that they would all be in operation at the same time, but several may well be, and therefore something will be starved of water and the flow rate will drop dramatically.

Second, a combination boiler is a priority system, which means that when it is providing the hot water to the hot taps and other outlets it does not supply the heating system. In other words, the boiler gives priority to the domestic hot water when in operation; it does not do both heating and hot water at the same time. So, for example, in a home with say six people, every time the bath or shower is being run, or the washing machine requires hot water, or any hot tap is opened, the heating will not be on. As a result, you may find the radiators getting cooler on occasions.

Regular boiler and hot water storage cylinder

The advantages of having a stored vented domestic hot water supply are generally the opposite of the problems of the combi boiler and include:

* the water flow out from the taps will be good (this is not to be confused with pressure, as previously identified)
* the central heating is independent of the hot water (i.e. this is not a priority system)
* the kilowatt rating or output size of the boiler does not need to be as high
* there will still be a limited backup supply of hot water if the water mains supply is turned off.

The points above relate to a vented storage system. Note that if an unvented system is installed a large supply mains is still required to combat poor flow conditions (minimum 25 mm polyethylene). The disadvantages of the storage system are the opposite of the advantages of the combination boiler.

So, in conclusion, if there are only two or three people living in the property, the minimum pipe diameter is 22 mm and the occupants are prepared to wait a minute or two longer to run their bath, then a combination boiler might be a suitable system. Money will be saved on installation and on running costs.

However, where several occupants inhabit the home, creating a greater demand for hot water, it might be worth finding the space to incorporate a regular boiler and hot storage cylinder, preferably unvented, thereby ensuring good flow and pressure to all outlet points without disrupting the central heating demand.

3

domestic central heating

Within this final chapter you will discover the many different types of central heating systems to be found. As with the previous chapter on domestic hot water supply you will find there are many designs and again you will see that some take their water directly from the cold water mains supply, whereas others are supplied with water via a storage cistern. Central heating systems have undergone many changes in design since they were first employed using radiators and you will see how modern systems often use the actual building as the heat emitter to warm the occupant, with runs of pipework embedded within the structure of the building. The chapter concludes with a look at the various central heating controls that will be encountered and the ways of protecting the heating systems and pipework from possible damage.

There is a variety of different methods of domestic central heating, including:

* electric storage heaters
* warm-air heating
* under-floor heating (radiant heating)
* water-filled radiators.

Of the above, water-filled radiators represent by far the most common system and therefore will be the main focus of this book. Of the others:

* electric storage heaters use cheap-rate electricity at night to warm up heat-retaining blocks, designed to slowly release their heat throughout the day
* warm-air heating consists of a network of ducting to distribute pre-heated warm air around the home
* under-floor heating uses either heated electric cables or water-filled pipe coils to warm up the structure of the building.

More and more new developments are including under-floor heating as an alternative to the more traditional radiator system. Under-floor heating is referred to as radiant heating and merits a review of its design in order to understand how it works effectively, compared with water-filled radiator systems.

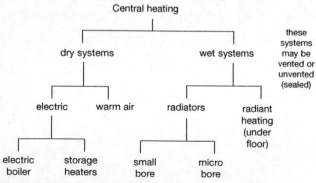

Figure 3.1 *Types of central heating systems.*

Radiant heating uses infrared heat rays that do not warm up the air through which they pass but the structure upon which they fall. In other words, radiant heating does not directly increase the temperature of the air in a room; instead, it warms up the structure of the building.

When a person enters a room their body tries to become the same temperature as the surrounding structure and, as a consequence, if the building is cooler than you are, infrared heat is lost from your body as it tries to even out the temperature difference. If, however, the structure of the building is warm, no heat will be lost from your body in this way. As a consequence, the ambient temperature of the room can in fact be cooler than your body and the building as the air temperature does not unduly affect your body temperature.

Coils filled with water are laid within the floors and, if they are left on long enough at a temperature of around 40°C, they will emit sufficient radiant heat to slowly warm up all the

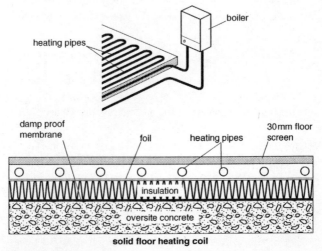

Figure 3.2 *Radiant heating.*

surfaces and solids within a room to a temperature compatible to that of the human body – around 33°C.

The advantages of having a radiant heating system include:

* cooler room temperatures which create a sense of freshness
* less transference of dust and airborne bacteria caused by the effects of convection currents
* very low water temperatures resulting in greater efficiency from the boiler (typically around 90 per cent – efficiency is explained later in this chapter).

Traditionally, UK homeowners only put on their heating for a few hours in the morning and a few hours in the evening. This limited amount of time is rarely sufficient to warm the whole building and, as a result, higher flow temperatures are used to warm the structure. This creates a certain amount of discomfort underfoot due to the elevated water temperature, and also reduces the efficiency of the boiler. For these systems to work really well, long periods of low-temperature water heating are required.

The other major disadvantage of this system is the problems created by a leak in the pipe coil. Fortunately leaks are quite rare, but it can prove very costly to find the leak and make the repair.

Central heating systems using radiators

Unlike under-floor heating, traditional water-filled radiators warm up the air surrounding the large metal surface of the radiator. It is this warming of the air that creates convection currents within the room. Convection currents are the flow of warm air around the room, caused by the hot air rising as it expands and becomes lighter, and the cooler, heavier air falling to replace the void – the cycle continues until the room is warm.

The pipe layout of this sort of central heating system can be of several designs, although around 95 per cent of all domestic heating systems using radiators use what is called the two-pipe system, i.e. there are just two pipes leaving the boiler. These two pipes, the flow and the return as they are called, travel around the house to the various radiators. At each radiator a tee connection is made to a pipe that branches off to feed a valve, usually found at one end of the radiator. The two pipes terminate at the final radiator.

For the past 50 years or so a circulating pump has been used to circulate the water around the heating system. Very rarely, in older properties, gravity circulation systems can still be found (Chapter 2). Sometimes these systems use solid fuel (wood or coal) and, unlike gas- or oil-burning appliances, you cannot simply switch off the flame, so a radiator or two is incorporated as a heat leak from the boiler, allowing heat to escape naturally from the boiler by gravity circulation. However, these systems are now quite antiquated and ought generally to be replaced.

Other central heating designs, such as the one-pipe circuit or the 'reversed return' system, can also be found, but due to their rarity in the domestic home they fall outside the scope of this book and have been omitted to avoid confusion. See Appendix: Taking it further, for further reading on these systems.

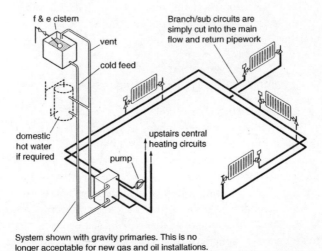

f & e cistern

vent

cold feed

Branch/sub circuits are simply cut into the main flow and return pipework

domestic hot water if required

upstairs central heating circuits

pump

System shown with gravity primaries. This is no longer acceptable for new gas and oil installations.

Figure 3.3 *Two-pipe system of central heating.*

The water to the system shown in Figure 3.3 is supplied via a feed and expansion (f & e) cistern found in the roof space (Chapter 2). This type of design is referred to as a vented system. However, the water may have been fed directly from the cold supply mains via a special filling point, in which case the system would be referred to as a sealed heating system and so would not be under the influence of atmospheric pressure.

Note also that the boiler is used to heat up the domestic hot water. In the system shown, a circulating pump is only used to force the water around the heating circuit. The water in the hot water cylinder circulates due to the effects of gravity (i.e. convection currents where the lighter hot water rises and heavier cold water sinks, as discussed in Chapter 2). This design does not comply with current Building Regulations but may, nevertheless, be the system that you have. Modern systems use a circulating pump to provide a more efficient system (a fully pumped system) as shown in Figures 3.4 and 3.8.

The installation of a modern central heating system fuelled by either gas or oil must comply with the latest edition of the Building Regulations. Systems that were installed prior to the current laws do not need to be updated, but should you replace your boiler at some time in the future you will need to upgrade your system as appropriate.

Sealed heating systems (closed systems)

What is a sealed heating system?

The term sealed system relates to systems that, once they have been filled up, usually via a temporary cold mains connection, have the temporary hose connection removed and the system closed off. The water is now trapped within the system and so it is not under the influence of atmospheric pressure.

Combination boilers are installed as sealed systems. They are designed with a temporary mains water filling connection to the central heating water and a permanent cold water mains supply for the domestic hot water draw off. The reason that the

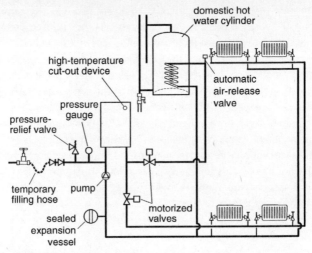

Figure 3.4 *Sealed heating system.*

temporary filling connection is disconnected from the water supply is that:

* it is a Water Regulations requirement
* chemicals may be added to the central heating pipework and if these are drawn back into water authority's water supply it would lead to contamination.

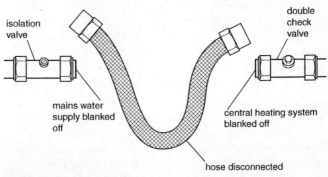

Figure 3.5 *Temporary filling loop.*

How are sealed systems different from vented or open systems?

The water in these systems is trapped within a closed circuit and therefore is not subject to the influences of atmospheric pressure. The expansion of the water, due to it being heated, is accommodated within a sealed expansion vessel. This expanding water creates additional pressures within the system and cause it to rise in excess of 1 bar pressure. In fact, these systems are invariably slightly pressurized, as a manufacturer's requirement upon filling to a typical pressure of 1.5 bar. As the pressure increases within the system so does the temperature at which the water boils. This could prove dangerous should excessive pressures develop, so the following safety controls need to be included at the time of installation:

* a temperature cut-off device, designed to shut down the appliance if the temperature exceeds 90°C
* a pressure-relief valve (safety valve) which can open to relieve the pressure from within the system if it becomes too great.

Sealed expansion vessel

In the case of vented systems of central heating, the water expansion resulting from the heating process is accommodated within the f & e cistern. Sealed systems, however, do not have this cistern open to the atmosphere, therefore the expanding water is taken up within a special steel container often found within the boiler casing itself.

The vessel contains a rubber diaphragm that separates it into two compartments. One side is filled with air to a pressure equal to that of the water in the system when it is cold; on the other side the system allows water to flow in and out as necessary. As the water heats up it expands and enters the vessel, pressing against the diaphragm and squeezing the air on the other side of the diaphragm into a smaller space, thus causing the pressure to increase. When the system cools, the increased air pressure forces the water back out into the system. Note that this expansion vessel is of a different design

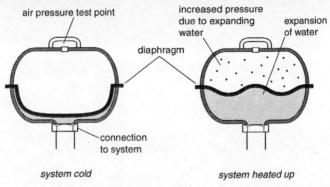

air pressure test point

increased pressure
due to expanding
water

expansion
of water

diaphragm

connection
to system

system cold

system heated up

Figure 3.6 *Operation of a sealed expansion vessel.*

from that used for a system of unvented domestic hot water, where a rubber bag is used to contain the expanding water (Figure 2.12).

Fully pumped systems and the location of the circulating pump

In a fully pumped system the circulating pump creates pressure within the pipework. It creates a positive, or pushing, force as the water is thrown forward from the pump and a negative, or sucking, force as it is drawn back into the pump when it returns from its journey around the system.

In the case of the sealed system (Figure 3.4), often the pump is incorporated within the boiler, installed on the pipe as it leaves the boiler. Because a sealed system is not subject to atmospheric pressure, half the system is subject to positive pressure and half to negative pressure. The pressure gradually reduces from the pushing force to zero, and the suction slowly gets stronger as the water returns to the pump. As a consequence, provided that there are no leaks, air cannot be drawn into the system.

This is not the case with a open-vented system. Figure 3.7 illustrates the principle that the cold feed enters the system at the point where the influence of the pump changes from

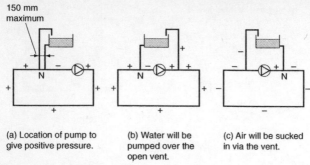

150 mm
maximum

(a) Location of pump to
give positive pressure.

(b) Water will be
pumped over the
open vent.

(c) Air will be sucked
in via the vent.

Figure 3.7 *The principles of correct pump location.*

positive to negative pressure. This point is referred to as the neutral point.

In Figure 3.7(a) the system is working well – the pump is creating positive pressure (above atmospheric pressure) around the whole system, which ensures that there are no micro-leaks (very small openings allowing the passage of air but not water) that will allow air to be drawn into the system. In this same system (3.7a), if the pump were installed the other way round it would create a negative pressure throughout (below atmospheric pressure). This would work fine, but air could be drawn in, for example, through radiator valve gland nuts, where the spindle turns (a typical micro-leak). Therefore, to ensure a good design always aim to get a positive pressure.

However, in the system in Figure 3.7(b) the open vent pipe is under a positive pressure and therefore will allow a quantity of water to discharge into the f & e cistern, subject to the head pressure created by the pump, and in so doing will oxygenate the water.

In the system in Figure 3.7(c) the open vent is subject to the negative pressure of the pump, so air will be drawn into the circulatory pipework. This configuration is often overlooked as it is not easy to spot. It can be identified by submerging the open vent in a cup of water – if it is sucking air into the system it will suck the water up from the cup.

When air is being drawn into your installation, not only is this inconvenient, causing radiators to fill with air and preventing them from working correctly, but it is also slowly and surely corroding your system from within as the oxygen in the air, when mixed with water, causes the iron radiators rapidly to corrode and rust away. The key thing to check, with an open-vented fully pumped system, is that the open vent connection is within 150 mm of where the cold feed joins the circulatory pipework.

The air separator

Heating installers sometimes incorporate an air separator into the pipework to serve as the collection point for the cold feed and open vent pipe. This fitting ensures that the required close grouping of the cold feed and vent is maintained and also creates a situation where the water becomes shaken and turbulent as it flows through the fitting. This helps the air molecules that are within the water to dissipate and escape by forming bubbles and rising up out of the system through the open vent.

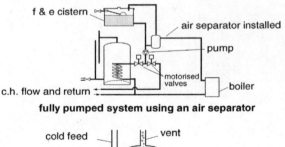

fully pumped system using an air separator

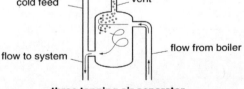

three tapping air separator
(the cold feed is introduced within 150 mm of vent)

Figure 3.8 *Use of an air separator.*

Micro-bore systems

Micro-bore is the name given to a central heating design that uses very narrow water pipes. At first sight, the pipe layout looks rather different but in fact it still follows the same basic design principles as the two-pipe system. Look carefully at the illustration of the micro-bore system in Figure 3.9 and you will see that a flow and return connection has been run from the boiler to each radiator. The main difference between micro-bore systems and the usual systems using 22 mm and 15 mm pipework is that instead of using tee joints at the connection to each radiator, a manifold is employed, from which several branch connections are made. (In order to show another variation on the theme of central heating design, the micro-bore system shown in Figure 3.9 has been run from a combination boiler.)

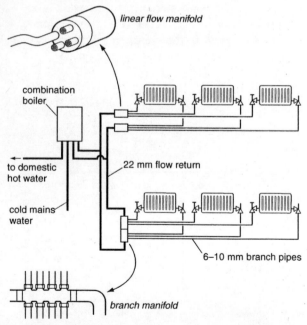

Figure 3.9 *Micro-bore heating system.*

Radiators and heat emitters

There are many different types of radiators, including modern fancy-shaped towel rails, skirting heaters, panel radiators, convector heaters, old-fashioned cast-iron sectional column radiators – call them what you like, but they all basically do the same job of warming the room in which they are installed. They warm the air in close contact with the radiator, and convection currents circulate the warm air around the room, as discussed earlier. Some designs are more effective than others: for example, the convector heater incorporates metal fins to assist in the distribution of the heat from the radiator to the air.

Manufacturers indicate the heat distribution from a particular heater as its kilowatt output, i.e. the higher the kilowatt output the greater the rate at which heat can be emitted. This must be taken into consideration when fitting heaters as it would be useless to install a radiator that is too small for a room because the occupants would feel insufficiently warm. Similarly, a radiator that was too large would occupy more wall space than necessary and would make the overall heating system less efficient. The room may warm up more quickly, but the amount of fuel used to heat up the larger volume of water contained within the radiator would increase.

The size of heater for a particular room can be calculated, but this requires the use of special tables and calculations taking it beyond the scope of this book. The process is not, however, particularly complicated and those who are interested in learning more should see the further reading suggestions in Appendix: Taking it further.

Radiator valves

A control valve will be fitted to each end of your radiator:
* one is designed to open and close the radiator as required
* the other valve, referred to as a lockshield valve, is non-adjustable and will have a plastic dome-shaped cap.

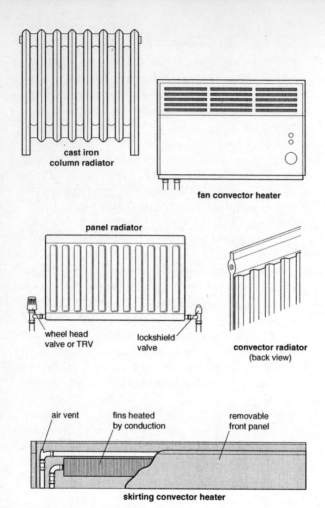

Figure 3.10 *Types of radiators and heaters.*

The first valve, used to open and close the radiator, may be a simple plastic-headed on/off control valve or a thermostatic radiator valve (TRV). For many years it has been the heating system installer who has chosen whether to use a manual valve

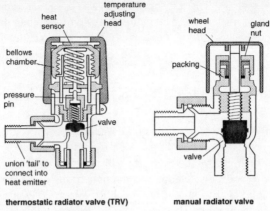

Figure 3.11 *Radiator valves.*

or a TRV, but current Building Regulations dictate the use of
TRVs. The only radiators that can be fitted with a manual valve
are those connected to radiators in rooms that have a room
thermostat.

The TRV automatically closes off the water supply to the
radiator when the room has reached the desired temperature
and therefore saves on the amount of fuel by avoiding
continuously supplying heat to a sufficiently heated room.

The lockshield valve, located at the other end of the
radiator, has a specific purpose in that it controls the amount
of water that flows through the radiator. It is identical to the
manual on/off valve except that the plastic head does not have
an internal square socket to fit over the turning spindle of the
valve. This valve has been pre-adjusted by the installer with a
spanner at the time of installation when balancing the system.

Balancing

In order to ensure that the first radiator on the heating
circuit does not take all of the hot water flow from the boiler,
due to the water taking the shortest route through this first
heater rather than going around the whole system, the lockshield

valve is partially closed. By having this valve open by, say, only half a turn, most of the water is forced to continue along the heating circuit to the next radiator. Further radiator lockshield valves are also adjusted as required to force the flow of water throughout the whole system.

Knowing which is the first and which is last radiator in the system

Basically, when you turn on the heat source of a cold system, the first radiator to get hot is the first, or nearest to the boiler, and will have the shortest circuit, the next to heat up will be the second radiator and so on throughout the system.

Note: If you ever need to turn off the lockshield valve with a spanner, for example when removing the radiator for decorating purposes, remember to count the number of turns to close the valve, so that when you re-open this valve you open it by the same number of turns, otherwise you might find that some radiators on your system do not reach their desired temperature because you may have affected the balancing of the system.

You should also note that if you have a micro-bore system then sometimes both valves are found at one end. This is achieved by utilizing an internal tube in the radiator to distribute the water flow as necessary, as shown in Figure 3.12.

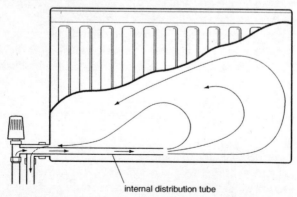

internal distribution tube

Figure 3.12 *Micro-bore connections at one end of a radiator.*

Air within the system and air vents

Prior to water entering the central heating system air will be inside. As water enters the system of pipework the air will be trapped in high pockets and, as a consequence, will prevent the system from operating correctly. This air is expelled from any high points such as the tops of radiators by small openings into which air-release valves have been installed.

The installer of the system will aim to run the pipework in such a way as to avoid trapping air. Where this is unavoidable, an automatic air-release valve can be inserted in the pipeline. This device contains a small float with a valve attached to its top end, so if water is present the float rises and the valve blocks up the outlet, whereas if there is no water within, the float drops and opens its outlet point or vent hole.

In addition to letting air out of the pipework and radiators, it is also necessary to open any air-release points when the system is being drained down, otherwise it will take forever to empty as air needs to enter the system in order to facilitate the removal of the water.

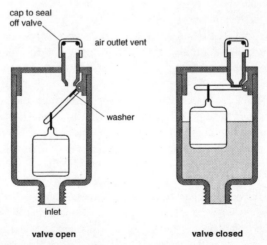

Figure 3.13 *Automatic air-release valve.*

The boiler

What about the heat source for the system? In its most fundamental form this is simply a metal box that is surrounded by a fire. In fact, the first heating systems were just this, a metal box referred to as a back boiler, found within the fireplace of the lounge. Surprisingly, there are a few still kicking around out there in some older properties.

Boilers today are fully automatic devices that turn up the heat as necessary and, with the exception of solid fuel systems, completely turn off when not required. The word 'boiler' is used but in fact the water never actually boils within the appliance, or in any case it should not, otherwise there would be something drastically wrong. The water is just heated until the required temperature is achieved, as set by its built-in thermostat, and then the heat source turns off. The fuels that could be used for the boiler include:

* solid fuel, including coal, wood and straw
* electricity
* gas
* oil.

Electric boilers are quite rare and so fall beyond the scope of this book. The remaining fuel types, however, have been used in boilers for many years, and the design of the boiler has developed into a very efficient appliance, unlike those of yesteryear.

Solid fuel has limitations in its design, and because these boilers tend to be more labour intensive – i.e. you need to load the fuel and empty the ash – they are not very popular and account for around only 0.5 per cent of all installations. Around 92 per cent of installations use gas and the rest use oil.

Due to developments over the years, there are many different boiler designs from many different manufacturers, with a never-ending list of models applicable to the particular designs. But fundamentally they all fall into one of four basic types:

* natural draught open-flued
* forced draught open-flued (fan-assisted)
* natural draught room-sealed
* forced draught room-sealed (fan-assisted).

So what do all of these names mean? Essentially, they relate to the method by which air is supplied to the boiler.

* Natural draught or forced draught indicates whether or not the appliance has a fan incorporated to assist in the removal of the combustion products to the outside.
* Open-flued boilers take their air from within the room where the boiler is located.
* Room-sealed means that the air is taken from outside the building's walls.

Figure 3.14 provides an illustration of these four designs.

The boiler in your home will be of one of these designs. For example:

* If you have a back boiler located behind a gas fire in the lounge, you have a natural draught open-flued boiler.
* If you have a large freestanding boiler in your kitchen, with a flue pipe coming from the top, travelling into a chimney or passing through a pipe to discharge up above the roof, again this is likely to be a natural draught open-flued boiler.

Both of these boilers take their air from the room in which they are installed, and this air is replaced via an air vent to the outside.

If your boiler has a terminal fitting flush with the wall it is most likely to be a room-sealed appliance.

* If this terminal is quite large it will be of the natural draught type.
* If it is smaller, say about 100 mm in diameter, it will be fan assisted.

These boilers do not take the air required for the combustion process from the room, but directly from outside.

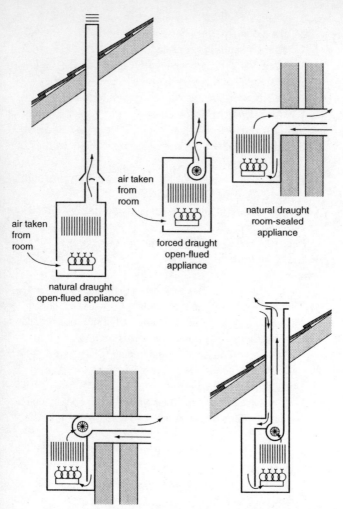

air taken from room

air taken from room

natural draught open-flued appliance

forced draught open-flued appliance

natural draught room-sealed appliance

forced draught room-sealed appliances

Figure 3.14 *Boiler designs.*

There are many variations of boiler design, dictating the location of the fan or the route of the flue pipe – which may be vertically through a roof or horizontally out through the wall – but they all fall within the four basic types listed above.

In addition to the basic boiler designs, boilers are further classified into four generic types:

* non-condensing regular boiler
* non-condensing combination boiler
* condensing regular boiler
* condensing combination boiler.

The differences between regular boilers and combination boilers have already been discussed in Chapter 2, but a new term is used here: 'condensing'.

What is a condensing boiler?

This is a boiler designed to take as much heat from the fuel and combustion products as possible and, as a result, is much more efficient. It is sometimes referred to as a high-efficiency boiler.

All domestic boilers installed prior to 1988 were designed in such a way that no consideration was given to the heat contained within the combustion products that were discharged from the boiler. If you were to take a thermometer and measure the temperature of the flue gases as they left the terminal, you would get a reading of something like 160°C. This is clearly a waste of heat and therefore of fuel. The condensing boiler is designed so that these combustion products are cooled to as low a temperature as possible, thereby using as much of their heat energy as possible.

For the traditional central heating system using radiators, this flue temperature would be somewhere around 80°C. This temperature could be reduced even further to, say, 45–50°C where a radiant under-floor heating system was installed (as discussed earlier). Where the appliance reduces the flue products down to a temperature of less than 54°C – i.e. the dew point of water – water that is produced as part of the combustion process condenses and collects within the boiler and is subsequently drained from the appliance.

These boilers, when in operation, especially when it is cold outside, are easily identified by the water vapour as a mist, referred to as a 'plume', discharging from the boiler terminal.

How do these boilers extract all this extra heat? Basically, the boiler has a larger and more tightly grouped heat exchanger or, in some designs, such as the one illustrated in Figure 3.15, has a second heat exchanger through which the flue products pass. The heat exchanger is the part that contains the central heating water over which the hot products of combustion pass.

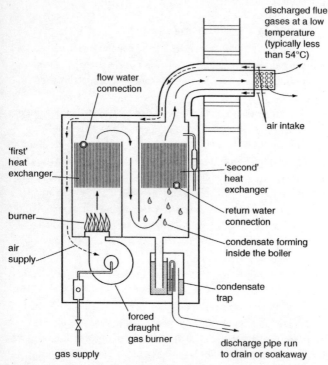

Figure 3.15 *Internal view of a condensing boiler.*

High-efficiency boilers (HE boiler)

Of the boiler designs identified, those that work to the highest standard are of the forced draught room-sealed type. A modern boiler has electronic ignition and a highly efficient heat exchanger, making it far superior to the old cast-iron boilers installed 30 years ago that operated on gas with a permanent pilot flame acting as the ignition source for the boiler. These old boilers might be operating at about only 50–60 per cent efficiency, whereas modern boilers could be operating at efficiencies of over 90 per cent.

When talking of efficiencies, one is effectively talking of the running cost. For example, for every £100 spent on fuel, if your boiler is only 55 per cent efficient you will be getting only £55 worth of heat, and £45 would simply be going up the chimney. But where your boiler is 90 per cent efficient you will be getting £90 worth of heat for every £100 spent.

It is because older boiler designs waste fuel in this way that current regulations no longer permit them to be installed. If you need a new boiler the chances are, with a few exceptions, that the heating installer will be bound by law to install a boiler with an efficiency of 86 per cent or higher.

Heating controls

In your home you may or may not have all of the controls listed below; in fact, you may have no more than a switch to turn the power on to the boiler and pump. However, the design of a modern central heating system will use a whole range of controls in order to provide an efficient system. One requirement of the current Building Regulations for all new and replacement systems using gas or oil as the fuel source is to have a minimum of the following controls:

* full programmer or independent time switching for heating and hot water
* room thermostat, providing boiler interlock

* cylinder thermostat (where applicable), providing boiler interlock
* TRVs on all radiators, except in rooms containing a room thermostat
* automatic bypass valve (if necessary).

These controls all serve the function of reducing the amount of fuel required to heat the water and thereby increasing the efficiency of the system. In other words, they save fuel. Should you need to undertake any major renewal work in your home, such as replacing the boiler, your system controls will need to be upgraded as necessary and include all of the controls listed above.

With the exception of the TRVs previously identified, what do each of the remaining controls do?

Full programmer

This is in effect a fancy clock. It allows the heating to come on at specific times as set by the occupant of the building. Modern installations require the use of what is referred to as a full programmer. This basically means that the heating circuit(s) and domestic hot water circuit can be controlled independently, allowing separate time settings for the heating and the hot water. Earlier designs of programmers did not have this independence, for example:

* mini-programmers allowed heating and hot water to be on together, or hot water only (but not heating only)
* standard programmers allowed heating and hot water to be on on their own, but used the same time settings.

These older time controllers will need to be replaced if the boiler is replaced, thereby complying with the current Building Regulations.

Room thermostat

A room thermostat is a device that senses the temperature of the room. When the temperature as set by the occupant is reached, an electrical contact is broken inside the thermostat to switch off the electrical supply to the pump or motorized valve found on the pipe serving the heating circuit. With no electrical supply the water ceases to be pumped around the circuit.

The room thermostat is normally positioned in a living room/lounge at a typical height of about 1.5 metres, but not in a position where it will be affected by draughts or by heat from the sun shining through a window. The thermostat should not be located in a room where there is an additional heater, such as an electric or gas fire – the hall might be a good alternative.

It is essential that the room selected for the thermostat does not have a TRV fitted to the radiator within the room because if the TRV closes, the room thermostat will fail to reach its operating temperature, and the heating will be on constantly. The idea of incorporating the room thermostat is to close off the heating circuit when the desired temperature has been reached in the living room. If the room thermostat is off, provided that the cylinder thermostat is not activated, the boiler and pump will be turned off, thereby saving fuel.

Some older systems may not have a room thermostat and just have TRVs fitted to the radiators to control the flow. These systems would need to be upgraded should the boiler be replaced at some time in the future.

Cylinder thermostat

The cylinder thermostat is a device fixed to the side of the hot water cylinder, about one-third up from the base. It is set by the installer so that it activates when the top of the hot water cylinder has reached a temperature of around 60°C. As with the room thermostat, when the desired temperature is achieved the electrical contact is broken inside the unit, which switches off the electrical supply to the motorized valve on the pipe circuit to the cylinder heat exchanger coil. Older systems may not have a cylinder thermostat. This is a situation that would need to be rectified should the boiler or cylinder be replaced, bringing the system in line with the mandatory regulations now in force.

Boiler interlock

Boiler interlock is when the boiler is linked with the thermostat system so that the boiler will only ignite if heat is required by either the domestic hot water or the central heating system, as regulated by the cylinder and room thermostats respectively.

Older systems did not always have a room or cylinder thermostat. For example, central heating systems were often designed only with TRVs fitted to the radiators, and gravity circulation of hot water to the cylinder from the boiler was allowed to continue until the boiler thermostat was satisfied.

Sometimes, to prevent the domestic hot water becoming too hot, a mechanical thermostat was installed in the return pipe to close off the flow of water in the circuit, and the boiler thermostat was the only control for switching the boiler on or off. Invariably it continued to heat up and cool down, night and day, as the boiler slowly lost its heat to the surrounding atmosphere. This is referred to as 'short cycling' and is clearly a drastic waste of heat and fuel, and this is what boiler interlock prevents.

Systems without boiler interlock need to be upgraded when major work is undertaken on the system, such as when replacing the boiler. Where you only intend to replace the hot water cylinder you must include a cylinder thermostat to operate a motorized valve to close off the circuit and switch off the boiler, but you do not need to upgrade the central heating controls. However, if you replace the boiler, both the cylinder thermostat and the room thermostat must be provided, thereby providing total boiler interlock.

Automatic bypass valve

This device is a valve, fitted in a pipeline, that opens automatically to allow water to pass. There are several reasons why they are sometimes incorporated in the pipe circuit, such as because the boiler has a pump overrun facility. This facility is needed in systems where the pump must continue running for a time after the boiler has switched off in order to allow the heat within the boiler to dissipate and for it to cool down sufficiently, thereby preventing heat damage to the boiler itself.

If the motorized valves of the central heating circuit and domestic hot water circuit are open they will allow the water to flow, but where these are closed, due to the temperatures of their circuits being satisfied, there will be nowhere for the water to flow. As a result, pressure will build up within the flow pipe from the boiler and this will press against the spring-assisted valve of the automatic bypass to force the valve to open. Some boilers come with a pre-installed automatic bypass.

Prior to the automatic bypass, a slightly opened manually set lockshield valve was installed by the plumber, but this method is no longer permitted to serve this function as it can reduce the efficiency of the system.

Motorized valves

Older central heating systems will not have these controls because, prior to the 1980s, systems generally were installed as shown in Figure 3.3. These older systems either had TRVs fitted to all but one radiator on the system to control the room temperature, or a room thermostat was used to control the heating requirements, which switched off the pump when the temperature within the room where the thermostat was located reached the required level. The temperature of the domestic hot water was generally only regulated by the boiler thermostat. These earlier systems, of which many thousands are still in existence, are far less efficient than the modern well-designed systems that use a motorized valve to close off the water supply to a particular circuit.

Closing off the motorized valve by way of the electrical power supply, from the room or cylinder thermostat as appropriate, provides a situation where the boiler is prevented from firing unnecessarily. The boiler of the modern system cannot fire unless either the room or cylinder thermostat is

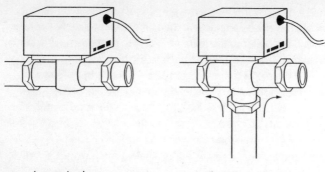

two-port valve   three-port valve

Figure 3.16 *Motorized valves.*

calling for heat, as it is these controls that send the power supply
to feed a motorized valve.

The motorized valve itself consists of a small motor
positioned on top of a housing inside which a ball-shaped valve
is moved by the motor, opening or closing the route through
which the central heating or domestic hot water can pass. There
are two basic designs of motorized valve:

* two-port (zone valve)
* three-port (either mid-position or diverter valve).

When power is supplied to the motor in a two-port valve,
it turns and causes the pipeline to open. As the valve opens it
makes the switch contact inside the unit to allow electricity
to flow to the boiler and pump. Should the power to this
control be switched off the valve closes, assisted by a spring,
and in so doing breaks the electrical contact to the boiler
and pump.

There are two basic types of three-port valve: the diverter
valve and the mid-position valve. The older designed diverter
valve allowed the water to flow either from the central inlet
port to the outlet pipe feeding the domestic hot water circuit or
to the central heating circuit. In effect, it opened one route but
closed the other, i.e. diverted the water flow, hence its name.
This system was wired up to give priority to the domestic hot

water in the cylinder, so that while this was being supplied with heat from the boiler the central heating system had to be off. This was affected by an internal ball which pivoted on a fulcrum between the two outlet ports.

The second type of three-port valve, referred to as a mid-position valve, allowed the internal ball valve to stop in the mid position as it was swinging across to close off one of the outlets, thus allowing water to flow to both the heating and domestic hot water at the same time, should both the room and cylinder thermostats be calling for heat. This mid-position valve therefore had the advantage over the earlier diverter valve. It must be understood, however, that while the valve is in the mid position the amount of water flow that can be expected through the valve is restricted, so they are only suitable for systems that are not too large.

appendix: legislation

Gone are the days when everyone could do what they liked. Today there is a whole range of legislation affecting what we can and cannot do in our homes. There is no restriction on what you can do yourself, but you still need to ensure that your work is in compliance with the law.

Much of the work completed these days requires the issue of a completion certificate, which is something, incidentally, which you must insist upon when employing someone else to do the work for you.

* Do not assume that they are registered with a specific body (see below).
* Do not be fobbed off with, 'It's not applicable to what we are doing'.

Most activities completed these days require some form of certification. When you have the work done you may not care too much whether or not a certificate is issued, but:

* when you come to sell your home it may be picked up by the surveyor and prove costly to certify this work at a later date
* you may not be covered by your insurance should you wish to make a claim.

Work requiring notification under local building control

In all of the following situations you will need to notify the local building control officer of the local authority of any work carried out. This is in order to comply with the requirements of the Building Regulations.

Where you use a contractor to do the work, they may be registered with a validating body such as Gas Register[1], NAPIT[2], APHC[3] or OFTEC[4], which allows them to self-certify the work (note that there are other bodies). However, you need to check that they really are registered with a certificating organization or you may not receive the certificate for the work completed, as required by law.

* Drainage alterations
* Heating and hot water requirements
* Electrical systems
* Gas systems
* Oil installations
* Ventilation

A couple of points to note:

* Advice relating to any of the above can be sought simply and quickly by phoning your local council offices and asking to speak to the building control officer. The plumber you engage should be fully conversant with these rules, but I suggest you do not bank on this.
* Without certification you may have to remove what you install if it is discovered by the local authority to have been carried out without approval.

Requirements needed by operatives:

* For all work relating to gas installations, unless DIY, operatives must be registered with the Gas Safe Register.
* For all work relating to oil, operatives should be registered with OFTEC.
* For all work relating to drainage, operatives should be registered with an approved body, such as those listed previously or another such organization.
* For all work relating to the hot or cold water pipework, operatives should be registered with a water authority.
* For all work relating to electrical installation, operatives should be registered with an approved body, such as those listed previously or another such organization.
* For all work relating to the ventilation, operatives should be registered with an approved body, such those listed previously or another such organization.

Where the operative is not registered with a specific body the work can still be completed, but you may need first to seek approval in writing from the local building or water authority, as necessary.

taking it further

Further reading

Treloar, R.D. (2006) *Plumbing* (third edition), London: Blackwell Publishing.

Treloar, R.D. (2009) *Plumbing Encyclopaedia* (fourth edition), London: Blackwell Publishing.

Treloar, R.D. (2010) *Gas Installation Technology* (second edition), London: Blackwell Publishing.

Plumbing trade and professional bodies

Listed here is a selection of organizations from which the contact details of qualified operatives can be sought. The companies listed with organizations such as these will need to follow strict guidelines as laid down by the organization making the recommendation and, as such, the organization will be held accountable to some extent for the work that they undertake.

Association of Plumbing and Heating Contractors (APHC)

024 7647 0627   www.competentpersonsscheme.co.uk/consumers

Chartered Institute of Plumbing and Heating Engineering (CIPHE)

01708 472791   www.ciphe.org.uk

Gas Safe Register

0800 408 5500   www.GasSafeRegister.co.uk

Oil Firing Technical Association (OFTEC)

0845 65 85 080   www.oftec.org

In addition, the following sponsored government website is very useful: www.trustmark.org.uk